Bibliographic information published by the German National Library:

The German National Library lists this publication in the National Bibliography; detailed bibliographic data are available on the Internet at http://dnb.dnb.de .

Imprint:

Copyright © 2016 GRIN Verlag, Open Publishing GmbH
Print and binding: Books on Demand GmbH, Norderstedt Germany
ISBN: 9783668585218

This book at GRIN:

http://www.grin.com/en/e-book/381348/the-philosophy-of-chemistry

Patrick Kimuyu

The Philosophy of Chemistry

GRIN Publishing

GRIN - Your knowledge has value

Since its foundation in 1998, GRIN has specialized in publishing academic texts by students, college teachers and other academics as e-book and printed book. The website www.grin.com is an ideal platform for presenting term papers, final papers, scientific essays, dissertations and specialist books.

Visit us on the internet:

http://www.grin.com/

http://www.facebook.com/grincom

http://www.twitter.com/grin_com

The Philosophy of Chemistry

Name: Patrick K. Kimuyu

Philosophy of chemistry is a sub-branch of the philosophy of science. It is a new field that was hived from the traditional philosophy of science. It has acquired autonomy from the philosophy of physics under which it was regarded as a part. Its late evolution was due the assumption that most philosophers and scientists made in regard to the relationship between physics and chemistry (Justi & Gilbert, 2002). The assumption was that the chemistry can be reduced to physics. Most scholars in the philosophy of science argued that physics, under the principle of quantum mechanics, is the science that describes reality at its best, whereas chemistry, as the phenomenological science describe phenomenon as they are seen by human beings (Lombardi & Labarca, 2007). This paper discuses the philosophy of chemistry.

Approached from this perspective, chemistry is only a part of the physics discipline and does not have problems that need philosophical analysis. The advocates of this perspective argued that any philosophical problems appertaining to chemistry are essentially belonging to philosophy of physics. Therefore, the philosophical problems concerning quantum mechanics were discussed by philosophers of science (Gabbay et al., 2011). However, in mid 1990s the interest in the philosophy of chemistry began to engage the minds of philosophers, who increasingly questioned the traditional assumptions on the relationship between physics and chemistry. Today, many scholars are convinced that chemistry can not be reduced to physics. However, although philosophers of chemistry strongly argue in support of the opinion, many in the scientific community still hand on to the reductionist position (Schummer, 2010). The reductionist approach has led to the tendency to explain some issues of chemistry, for instance, the atomic structure, through physics principles. The reason chemistry is considered as a branch of physics is because it deals with particular processes that, however, be explained through quantum theory. The net effect is that physics is placed at the top of the hierarchy while chemistry is positioned at the bottom as an appendage of physics (Lombardi & Labarca, 2007). However, chemical philosophers have established the need to introduce philosophical discussions intended at establishing chemistry as an independent discipline free from the fundamentalism of physics.

2

Some scholars have argued that the issue of reduction is at the heart of the understanding of the philosophy of chemistry. They posit that it is essential to distinguish between the two views of reduction. These are the ontological and epistemological reduction. Ontological is the study of metaphysics and concerns itself with reality, its structure, and the components in it. Epistemology, on the other hand, deals with knowledge, its scope and limits. Consequently, ontological reduction makes reference to the ontological dependence of the properties and regularities of a part of reality upon the properties, entities, and regularities of another stratum that is considered to be ontologically fundamental. Consequently, ontological reductionism can be construed as a metaphysical thesis that displays the ontological priority of a given level of reality on which other levels of realities indirectly or directly reduce. Epistemological reductionism is a thesis upon which science should be unified (Schummer, 2010).

In the last two decades there have been efforts to liberate chemistry from the shackles of physical thoughts. These scholars have defended chemistry as an autonomous discipline on historical grounds citing the different historical traditions that saw the evolution of physics and chemistry (Ebbing & Gammon, 2005). The scholars argue that the epistemological reduction of chemistry to physics is not possible. The philosophical question is, therefore, raised, and calls for a philosophical discussion to address it (Lombardi & Labarca, 2007). Although the philosophy of science has not given much attention to the philosophy of chemistry before the 1990s, scholars opine that the historiography of philosophy of chemistry simply ignored what earlier philosophers had commented about chemistry. Schummer (2010) observes that the philosophy of science in the communist world had been broad enough to accommodate chemistry, especially between the years 1950s to 1900 (Schummer, 2010). The communist approach to the philosophy of science drew on the dialectical materialism promoted by Engels. According to Engel, chemistry was a facet of French materialism or mechanical materialism. For the chemical, mechanical and physiological level, he saw different types of movement each following its law as well as dialectical laws that would lead the transformation to higher levels (Lombardi & Labarca, 2007). The twentieth philosophers saw some merits in Engels theorization and expanded on it. They recognized that the

3

chemical phenomena could be used to illustrate the universal laws as advocated by Engels doctrine. For example, the philosophers saw the acid-base reactions could be used to exemplify Engels law of contradictions in regard to counter-acting forces that are found in nature (Schummer, 2010). The communist philosophers had an established position in tertiary science education so they could be committed to interpreting specific scientific facts, developments or problems in the framework of dialectical materialism. The philosophers were free to deal with chemistry as an autonomous field because Engel in his handling of the chemistry had reserved an own kind of movement for chemistry (Lombardi & Labarca, 2007).

In the western countries, the professional philosophers did not give the philosophy of chemistry much thought; however, scholars from other disciplines approached it albeit from their own perspectives and with specific issues. Scholars of chemistry education did recognize the need to reflect on the methodology and worked on the clarification of concepts (Brakel, 2000). Working chemists also did come across some philosophical questions when they were challenged by their research to reflect either on received notions or methodological ideas. For instance, some chemists working on isotope reflected on the theory of chemical elements. Others reflected on the notion of causation prompted by their studies on chemical catalysis. Theoretical chemists, working on the development of quantum models for chemicals, started questioning the reductionist perspective (Görs et al., 2005). Critics have observed that in the western world, there were no distinction between history and philosophy of science; consequently, historian of chemistry approached the field of philosophy of chemistry through philosophical concerns of the past; such as the metaphysical problem of atomism and the methodological problem of conceptual change as exemplified by such scholars as Kuhn in his chemical revolution (Görs et al., 2005).

A new discourse on philosophy of chemistry has been opened with the realization that the chemical environment is the key to most aspects of humanity well being. Chemists have devised processes used in the production of goods that are accepted as contributing to the conditions of life (Gilbert et al., 2003). Chemistry can also contribute to repairing the damage caused to the environment

4

by the exploitation of chemistry in the past. The philosophical question to ask is whether chemists should be expected to scrutinize the projects they undertake to the effects the project achievement could have on the practitioners and other people. Clearly, there are many questions that can be philosophically explored in the study of chemical philosophy (Brakel, 2000). The observation is buttressed by Stephen Toulmin in his book Human Understanding where he asserted that, Men displays their rationality, not through ordering their concepts in tidy formal structures, but through their readiness to respond to new situations with their open minds (Schummer 2010, p. 189). Hence, chemists engage in philosophical musing by questioning their activities and seeking clarification on some aspects of their work that are not too clear. Consequently, the philosophy of chemistry needs to define the research and international programs so as to make intellectual progress in regard to the nature of science, humanity, and human knowledge (Gilbert et al., 2003).

Chemistry and physics as different branches of science have similarities. Historically, chemistry dealt with what is today referred to as inorganic chemistry that is, the chemistry of substances not associated with living things. There were considerable analysis done to discover the existence of the elements, how they relate, and how they make the compounds found on the earth. The early works of chemistry were important to physics. For instance, the theory of atoms was substantiated through experiments in chemistry (Gabbay et al., 2006). The theory of chemistry, the chemical reactions themselves, was explained in the periodic chart. The chart brings out the strange relationships between the various elements and explains the rules as to which element is combined with which, and how it is combined. The rules were finally explained through the principle of quantum mechanics, making theoretical chemistry as physics (Talanquer, 2010). Scholars have opined that it is difficult to predict what happens in any given chemical reaction; however, any theoretical chemistry has to end up in quantum mechanics (Tsaparlis, 2003). The second branch of chemistry is called organic chemistry, or the chemistry that studies the elements or substances associated with living things. For some time it was believed that the organic substances could not be made by hand or from inorganic materials. It has,

5

however, been shown that the substances are the same as those made in inorganic materials, but exhibit a more complicated arrangements of the atoms (Brakel, 2000).

However, the area where the two sciences converge in their philosophy is in the principle of reduction. Physics follows the ontological reductionism when it shows that the principle of entities, properties and processes at a particular level are manifestation of the same at another level (Scerri, 2008). For instance, a solid macro property simply consists of the micro atomic properties and the relationship between the two is one of supervenience. It means that the macro properties are dependent on the micro properties. Just like in philosophy of chemistry, philosophy of physics uses epistemology reductions to explain or account for the ontological relations (Morrison, n. d.). Much of what is written on reduction dwells on the examination of the possibility of reducing one theory to another, for instance, the much-discussed issues of statistical mechanics and thermodynamics. In carrying the exploration of the epistemological reduction, another notion is the derivation of a set of laws from other main ones (Sjöström, 2007). The idea is to have a convergence of explanations derived from a common source where scientific generalizations are used to explain others, and consequently, offer a sense of scientific direction. The principle of grand reductionism says that scientific theories can all be traced to a few universal laws (Morrison, n. d.). What is significant is that reduction is investigated in terms of principles and not entities. All the sciences, such as chemistry and biology, agree that a good deal of scientific works involves the breaking down of the entities to their basic constituents (Baird et al., 2006). Thus, whereas the analysis gives substantial information on the physical world, it is also important to understand the ways the constituents behave and the law they follow as they make their complex systems. It is, therefore, clear that reductionism theory and methods are at the basis of the developed arena of modern science (Brakel, 2000). Its importance is to be found in physics biology and chemistry. Hence, classical mechanics gives the reductionist framework, whereas statistical mechanics offers the reconciliation thermodynamic law in explaining macroscopic properties through the microscopic components.

6

Conclusively, the philosophy of chemistry carried the assumption that chemistry can be reduced to physics. Most scholars in the philosophy of science argued that physics, under the principle of quantum mechanics, is the science that describes reality at its best, whereas chemistry, as the phenomenological science describe phenomenon as they are seen by human beings. Chemistry can, however, develop its line of philosophical questions on its own. Despite the need to develop the two sciences as different philosophical disciplines, they have areas where they converge, especially on scientific principles.

References

Baird, D., Scerri, E., & McIntyre, L. (2006). *Philosophy of Chemistry: Synthesis of a New Discipline,* New York, NY: Springer Science & Business Media.

Brakel, J. (2000). *Philosophy of Chemistry: Between the Manifest and the Scientific Image.* Leuven: Leuven University Press,

Ebbing, D., & Gammon, S. (2005). *General chemistry.* Boston, MA: Houghton Mifflin.

Erduran, S. (2009). Beyond Philosophical Confusion: Establishing the Role of Philosophy of Chemistry in Chemical Education Research. *Journal of Baltic Science Education,* 8, 5-14.

Gabbay, D., Thagard, P., Woods, J., & Hendry, R. (2011). *Philosophy of Chemistry.* New York, NY: Elsevier.

Gilbert, J., Jong, O., Justi, R. David, F., Treagust, D., & Driel, J. (2003). *Chemical Education: Towards Research-based Practice,* New York, NY: Springer Science & Business Media.

Görs, B., Psarros, N., & Ziche, P. (2005). *Wilhelm Ostwald at the Crossroads between Chemistry, Philosophy and Media Culture.* Leipzig: Leipziger Universitätsverlag.

Justi, R., & Gilbert, J. (2002). Philosophy of chemistry in university chemical education: The case of models and modeling. *Foundations of Chemistry,* 4, 213-240.

Lombardi, O., & Labarca, M. (2007). The philosophy of chemistry as a new resource for chemistry education: *Journal of Chemical Education, 84,* 187-192.

McIntyre, L. (2007). The philosophy of chemistry: ten years later, *Synthese, 155(3),* 291-292.

Morrison, M. (n. d.). *Emergence, Reduction, and Theoretical Principles*:

http://www.isnature.org/Files/Morrison_Emergence_Reduction_Fundamentalism.pdfhinking

Fundamentalism.

Scerri, E. (2008). *Collected Papers on Philosophy of Chemistry*. London, UK: Imperial College Press.

Schummer, J. (2010). *Philosophy of Chemistry in Philosophies of the Sciences: A Guide*, London, UK:

Blackwell Publishing Ltd.

Sjöström, J. (2007). The Discourse of Chemistry (and Beyond). *Hyle an international journal for the philosophy of chemistry*, 13(2), 83-97.

Talanquer, V. (2010). Macro, Submicro, and symbolic: The many faces of the chemistry "triplet".
International Journal of Science Education, *37*, 1-17.

Tsaparlis, G. (2003). Chemical Phenomena Versus Chemical Reactions: Do Students Make the

Connection? *Chemistry Education: Research and Practice*, *4*, 31-43.